BRISTOL

Images from the Transport Treasury archive

200 Years of Train Travel Since 1825

© Images: Transport Treasury or as credited. Design: The Transport Treasury 2025. Text: Jeffery Grayer.

ISBN 978-1-913251-95-6

First published in 2025 by Transport Treasury Publishing Ltd. 16 Highworth Close, High Wycombe, HP13 7PJ
Totem Publishing, an imprint of Transport Treasury Publishing.

The copyright holders hereby give notice that all rights to this work are reserved.
Aside from brief passages for the purpose of review, no part of this work may be reproduced,
copied by electronic or other means, or otherwise stored in any information storage and
retrieval system without written permission from the Publisher. This includes the illustrations
herein which shall remain the copyright of the copyright holder.

www.ttpublishing.co.uk

Printed in Tarxien, Malta by Gutenberg Press Ltd.

Front cover - This view of the approach road and frontage of Bristol's Temple Meads station dates from May 1979. A fine array of colourful cars of the period includes such stalwarts as the VW Beetle, BMC Mini and the Ford Cortina. Also of note is the Bristol bus, a green liveried single deck RE type probably operating the route to Clifton, wearing the National Bus Company logo of the double N symbol, the former Bristol Omnibus Company having become part of NBC in 1969. Part of the original station building can be seen to the left whilst on the far right the trainshed roof is just visible but occupying a central position is the stately frontage dating from the 1870s. *(Terry Nicholls)*

Frontispiece - Western class diesel hydraulic No D1072 *Western Glory* wearing BR blue livery takes pride of place in this view of the western end of Temple Meads as it departs southwards with a service for the south west. At the platform on the right is an unidentified "Peak" class diesel electric locomotive being observed by the usual crowd of platform end "spotters". D1072 was shedded at Bath Road depot from June 1969 until transfer to Plymouth's Laira shed in October 1971 from where it was withdrawn in November 1976. *(Terry Nicholls)*

Rear cover - The eastern end of Temple Meads seen in the early 1970s with a couple of Brush Type 4s on view a class then operating many principal passenger services. A relic of the steam era is still in evidence in the shape of the platform mounted water tank. Other items that will become extinct in the future are the Royal Mail trolleys for use with mailbags which are now a thing of the past as mail generally no longer goes by rail with the last vestige, a dedicated mail service using EMUs, having been withdrawn in September 2024. *(Author)*

Copies of most of the images within this book are available direct from The Transport Treasury. Please quote the book title, page number and image reference.

INTRODUCTION

On 27 September 1825 George Stephenson's steam powered "Locomotion No 1" travelled 26 miles between Shildon, Darlington and Stockton. The choice of Bristol to feature in this series of books marking the 200th. anniversary of railways in the UK in 2025 is in recognition of the fact that the city became, and continues to be, a major railway hub serving the South West of England. In its early days Temple Meads played host to three railway companies.

The Great Western Railway arrived in August 1840 and was accommodated in a magnificent Brunel building featuring a mock hammer beam roof which, when constructed, was the largest timber beam roof in the country. Next to arrive was the Bristol & Exeter Railway in 1841 initially using the Brunel terminus which was not particularly suitable for trains departing in a south westerly direction, as it involved reversals to gain access to and from the terminus. The B&ER suffered this inconvenience until 1845 when it was accommodated in a considerably less salubrious but more operationally efficient station, which was described by observers as "a simple makeshift structure" and "a cowshed" set at right angles to the GWR building. Finally the Bristol & Gloucester Railway, later the Midland Railway, began running into the GWR station from 1844, although it did open a small terminus of its own at St. Philips for local services in 1870.

To relieve congestion at both stations with the increase in traffic in the 1850s and 1860s, it was decided to construct a new joint facility on the site of the B&ER station which after delays formally opened in January 1878. Apart from various platform extensions this is basically the station we see today. Pride of place must go to the vast curved train shed roof with a span of 125 feet and 500 feet in length, although not far behind in grandeur is the central block and clock tower built in the perpendicular Gothic style. Another fine building, constructed by the B&ER as their headquarters, remains in the shape of the Jacobean style structure now known as Bristol & Exeter House. Following the national trend of plummeting passenger numbers during the Covid epidemic Temple Meads handled some 9.3m passengers in 2022/23, making it the 40th. busiest station in the UK. It is particularly important as an interchange point between north/south and east/west services and sees some 2,320 scheduled services per week.

Bristol, formerly served by a number of prestigious named expresses, is among just a handful of large cities that have not had a plethora of competing stations for, apart from the small MR terminus at St Philips, traffic has always been concentrated at Temple Meads. Manchester and Glasgow for example both had four central stations, whilst Leeds, Liverpool, Leicester, Plymouth and Nottingham had three and Birmingham, Sheffield, Edinburgh, Bradford and even Bath had two. Bristol also pioneered the "Parkway" station concept with the opening of Bristol Parkway on the northern outskirts of the city in 1972.

It is a difficult task to represent the railways of Bristol in the 80 pages available here but aspects covered in this book include images of the three outstanding architectural features of its main station, some of the named expresses which served the city, together with a look at the depots dealing with a wide variety of motive power and the freight yards. Images from the steam age together with first generation diesels and views of the contemporary scene all feature in this volume which additionally covers some of the once substantial, but now much reduced, suburban stations that served the greater Bristol area. However retrenchment is not the whole story for in recent years new stations have opened at Portway Park & Ride and at Ashley Down, whilst Filton Abbey Wood replaced the former Filton Junction station in 1996. The prospect of electrification reaching Temple Meads is also a hopeful sign for the long term future of Bristol in the national railway network.

We begin this collection of images of Bristol's railways in this bicentennial year of UK railways by going back almost a century to September 1929, the date of this image of LMS 4-4-0 No 526 entering Temple Meads. Dating from 1898 this was one of the forty members of designer Samuel Johnson's Class 60 locomotives built at Derby, being originally numbered 63 changing to 526 in 1907 and finally to 40526 in 1950. It was withdrawn in 1956 after completing nearly 58 years in service. The stern warning notice regarding crossing the lines except by means of the footbridge reflects the fact that the subway between platforms was not provided until the following year, when the station was enlarged under a Government scheme whereby loans were made available to carry out large public works to alleviate unemployment during the Depression. This subway boasted at one time baths, toilets and even hairdressing salons. *(Dr Ian Allen)*

Left Top: This view of the splendid perpendicular Gothic ornamental stone turreted exterior of the main station dates from 28 May 1961 – five thirty to be exact by the station clock. A fine collection of vehicles of the period including the ubiquitous Ford "Pop", Morris Minor and Ford Anglia, adorns the forecourt with a particularly stately "horseless carriage" of a bygone age in front of the main entrance. This frontage, one of the stunning architectural features that make Temple Meads such an attractive station, dates from the 1870s and once sported a spire above the clock tower which was unfortunately destroyed in an air raid of January 1941. *(Leslie Freeman)*

Left Bottom: Another of the outstanding features of Temple Meads station is illustrated in this 1949 view of the interior of the original Brunel station with its hammerbeam roof, in front of which is the Digby Wyatt extension featuring the wall mounted signalbox on the right. After being taken out of railway use in 1965 it became a car park, as illustrated later in this volume, before becoming a home for the Bristol Empire & Commonwealth Exhibition which closed in 2008. Today the original station functions as an ideal exhibition space hosting a wide variety of events. *(Arthur Mace)*

Right: Whilst this area of the station was still active our photographer took this image of Standard tank No 82040 and Black Five No 44766 on 19 October 1962. These platforms were generally used by trains of the LMR operating services to Bath Green Park and to the Midlands and the North via Mangotsfield. No 82040 carrying an 82F St. Philips Marsh shedplate may be heading a stopping service to Bath Green Park whilst the 4-6-0, allocated to Bescot (21B) at this time, is awaiting a return to the Birmingham area. *(R C Riley)*

Left: Parked by the wall mounted signalbox which carried the cast iron nameplate "Bristol Old Station Signal Box" in the Digby Wyatt extension is one of the former GWR's streamlined railcars number W28W, one of the 1940/41 built batch with 48 seats. These railcars were eminently suitable for lightly patronised lines. Four of the 38 cars built were destroyed by fire including one at nearby St. Anne's Park in April 1956. Three have survived into preservation. *(George Heiron)*

Right Top: 4F No 44135 shares the oldest part of the station with a DMU on 16 July 1963. Following the decline in passenger traffic in the 1960s the platforms in this original Brunel terminus became redundant, closing on 12 September 1965, the area becoming a car park from February 1966. The new Bristol panel MAS box, which opened in 1970 was constructed unfortunately blocking any future rail access to the old station. *(A E Bennett)*

Right Bottom: This view dating from June 1962 is looking eastwards and shows the dilapidated remains of the former LMS terminus of Bristol St. Philips. This short branch from the Midland main line at Barrow Road into central Bristol hosted a local service from Bath Green Park until September 1953 when services transferred to Temple Meads. Freight traffic to the Goods station here, known as Midland Road, ceased on 1 April 1967 and was prominently featured in the British Transport Films 1957 production "Fully Fitted Freight" which covered the journey of the 4:48pm Bristol to Leeds freight express. Midland services had also run from Bath Green Park to Clifton Down via Kingswood Junction and Ashley Hill junction, to the east of Montpelier, but these finished in March 1941 as a wartime economy measure never to be reinstated. *(Ben Brooksbank, Creative Commons Attribution Share-alike License 2.0)*

The third great architectural feature of Temple Meads was of course its magnificent curved trainshed some 500 feet in length with a span of 125 feet. Seen from platform 8 underneath its protective canopy 4-6-0 Castle class No 4078 *Pembroke Castle* is on one of the through roads with a couple of coaches in tow. The W H Smith bookstall is a prominent feature on the opposite platform No 9. No 4078 was a Bath Road shed allocation for a few months at the end of 1958 and the beginning of 1959, possibly helping to date this image. *(Arthur Mace)*

We now turn to some of the named expresses that served Bristol, beginning with perhaps the premier service "The Bristolian" which was introduced in 1935 and became the pinnacle of GWR high speed luxury travel. Initially worked by 4-6-0 "King" class locomotives, this example of which is believed to be No 6019 *King Henry V* seen here entering Temple Meads in 1955 with a smart chocolate and cream rake of coaches forming the down service. However such powerful motive power became the exception rather than the rule as it was realised that the smaller 4-6-0 "Castle" class were in fact better suited to the task of hauling this express. *(Arthur Mace)*

Illustrating the point regarding Castle haulage is No 5034 *Corfe Castle* seen here on Bath Road shed after arrival with the down service and still sporting "The Bristolian" headboard. In 1958 for example the non-stop run of this restaurant car express between Paddington and Bristol was accomplished in 1 hour and 45 minutes. Departing from London at 8:45am, arrival in Bristol was at 10:30am with the return service leaving Temple Meads at 4:30pm and reaching Paddington at 6:15pm. This scene reveals the numerous piles of ash which were an inevitable feature of the operation of a steam depot and to the far left can be seen the depot mobile crane attempting to clear some of this debris by loading it into the adjacent truck. *(Eric Sawford)*

Another named express which graced the platforms of Temple Meads was "The Cornishman" seen here headed by Castle class No 5089 *Westminster Abbey* in the early 1960s wearing the appropriate headboard and reporting number C35. Although the name had been used earlier by the GWR for a service from Paddington to Penzance "The Cornishman" seen here began life under BR in 1951 operating from Wolverhampton via Birmingham, Stratford-upon-Avon, Cheltenham and Gloucester to Bristol thence to Penzance with a portion being detached at Exeter for Kingswear. In 1958 the end to end journey from Wolverhampton to Penzance, not one to be undertaken by the faint hearted, took no less than 9 hours and 10 minutes. Later changes saw "The Cornishman" diverted away from the Stratford-upon-Avon route and extended to Derby, Sheffield, Bradford and Leeds. *(Arthur Mace)*

Left: An earlier manifestation of "The Cornishman" carrying reporting number 675 is headed by an unidentified Castle class locomotive and captured here negotiating the curve away from Temple Meads, heading up the Midland line with the up service from Penzance. An Ivatt tank is seen on the right. The gas holders situated in the appropriately named Gas Lane to the east of Temple Meads are prominent in the background. *(George Heiron)*

Right Top: Another of the named expresses was "The Merchant Venturer" introduced in 1951 in connection with the Festival of Britain and seen here at Bristol on 16 April 1955 headed by King Class No 6015 *King Richard III*. It is carrying the headboard and reporting number 142 and has just detached itself from the down service. The train from Paddington called only at Bath Spa and Temple Meads and after a locomotive change would terminate at Weston-super-Mare. *(Stephen Summerson)*

Right Bottom: A headboard of a rather different kind is that advertising the "Bath Festival" which in 1958 ran from 29 May until 7 June. Bath has held music, drama and arts festivals of different types and under various titles since 1930. The event held in 1955 was called the 'Bath May Festival' and although no event was held in 1956 or 1957 from 1958 until 1987 events were known as simply the 'Bath Festival". Here Castle class No 7024 *Powis Castle* has charge of the 6:10pm service from Bristol to Paddington on 30 May 1958. *(Leslie Freeman)*

Moving on to a couple of enthusiasts' specials which visited Bristol in steam days here we feature that grand old trouper No 3440 *City of Truro* hauling an RCTS special of 28 April 1957, the year in which this veteran had returned to service. The occasion was the RCTS "North Somerset" special which originated from Waterloo with Class N15 No 30453 *King Arthur* in charge as far as Reading where the 4-4-0 took over for the run to Bristol. Here a couple of Ivatt tanks took the participants on a tour of local lines including the Bristol harbour branch, Wrington on the former branch from Yatton and Burnham-on-sea on the S&D from Highbridge. Returning to Bristol the tour was handed back No 3440 this time in harness with No 5528 for a journey down to Frome via Radstock. From here a return to the capital at Paddington was accomplished via Westbury, Newbury and Reading. *(A E Bennett)*

A much sadder occasion was the Western Region's "Farewell to Steam" special of 27 November 1965 advertised as the "Last steam train from Paddington" and hauled by the "Last Castle class locomotive in BR service". It was headed over part of the itinerary from Paddington to Gloucester Eastgate via Temple Meads by No 7029 *Clun Castle*. The tour then proceeded to Cheltenham St. James behind Western diesel hydraulic No D1006 *Western Stalwart* which returned the tour to Gloucester Central. Here *Clun Castle* took over for the run to Swindon from where a return to Paddington was undertaken by a pair of Class 37s Nos. D6881 and D6882. It is seen here at Temple Meads prior to setting off for Gloucester at about 12:30pm. (Alec Swain)

Left: Our final example of a named train, the *Isambard Kingdom Brunel Special* as indicated by the headboard, records the naming of a locomotive, Brush Type 4 D1662, at a ceremony which took place at Temple Meads on 20 March 1965. On this seemingly rather damp day classmate D1661 was named *North Star* at Paddington by Ray Gunter MP whilst D1662 was named by the Lord Mayor of Bristol in the original Brunel building at Temple Meads. It is seen here at the head of the "Bristol Flyer" special organised by the RCTS from Paddington to Bristol and return. The brochure advertising the tour included the following – "This trip, arranged at the request of the Western Region of BR, represents a step forward into the entirely new field of diesel hauled society special trains, and is indeed the first such train to be run by any organisation". D1662 under its TOPS guise as 47484 has been preserved by the Pioneer Diesel Group at Wishaw. *(Alec Swain)*

Right Top: Displaying part of the large running in boards provided at Temple Meads, this view of the eastern end of the station reveals Stanier Jubilee No 45602 *British Honduras*, these days known as Belize. It is hopefully about to depart with an excursion to the Midlands although the number of passengers' heads looking out of the coach windows may indicate some frustration at a delay. This 4-6-0 had a spell allocated to Barrow Road shed, from 1951 to 1956, but ended its days in 1965 at Holbeck shed in Leeds. The drawing power of the DMU to Severn Beach is obviously insufficient to turn the attention of the young spotters (complete with that vital accoutrement the duffel bag) away from the much more interesting steam locomotive opposite. *(George Heiron)*

Right Bottom: An earlier incarnation of the Jubilee class is this view of No 5642 *Boscawen*, named after a Royal Navy admiral, seen here wearing LMS livery. This view taken in September 1936 predates the renumbering of this locomotive in May 1948 following Nationalisation. It carries shedplate 22A indicating Barrow Road (its allocation from the year after construction in 1934 until a move to Newton Heath in the 1950s) on one of the through roads at Temple Meads with some interesting clerestory stock in tow. At the other platform is Castle class No 5021 *Whittington Castle*. *(Dr Ian Allen)*

Our third Jubilee representing the Midland influence at Bristol is this stunning nocturnal shot of No 45605 *Cyprus* awaiting departure from Temple Meads on an unrecorded date. Barrow Road shed had an allocation of these 4-6-0s with, for example, 10 being shedded there in January 1960. No 45605 was withdrawn from Burton depot in March 1964. *(George Heiron)*

Our final view in this section portraying some of the motive power to be seen in and about the station reveals one of the Britannia Pacifics in the shape of No 70053, sadly bereft of its former nameplate *Moray Firth* and taking water before departure with an excursion from Weston-super-Mare to Birmingham on 3 August 1965. Note that colour light signals are in evidence at the platform end whilst in the goods sidings behind, semaphores still rule. No 70053 remained in service until withdrawal from Carlisle Kingmoor in April 1967. *(Alec Swain)*

Passing Bedminster signalbox and carrying reporting number 263, signifying the 8:00am express from Plymouth to Crewe, County class 4-6-0 No 1014 *County of Glamorgan* passes the 74 lever box which was taken out of commission in April 1970, the station having become unstaffed in September 1968. (George Heiron)

Coming down to earth with more mundane motive power, Hawksworth pannier tank No 9488 awaits departure time with the 1.30pm (SO) service to Frome via Radstock West on 28 May 1958. Following the introduction of increased bus services in the late 1950s the former reasonably well patronised passenger traffic was badly affected and notwithstanding the halving of the service in September 1958, BR claimed the line was losing over £18,000pa. Thus it was that the final passenger train ran on 31 October 1959 although freight continued to be carried for some years afterwards. Another of the large running-in boards is visible, this time in its entirety in this view. *(Leslie Freeman)*

Turning now to some of the diesel classes that could be seen at Temple Meads we have this view of Warship diesel hydraulic D845 *Sprightly* wearing green livery with a small yellow headcode panel emerging from underneath Bath Road bridge on 26 June 1963. The headcode 1M99 indicates a service from Plymouth to Liverpool which included in the rake the first vehicle, which is a Royal Mail coach with lineside catching apparatus stowed inboard under the watchful eye of a couple of staff. Condemned in October 1971 after just a decade in service this North British Bo-Bo locomotive was broken up at Swindon the following year. The office block of Bath Road diesel depot constructed following the closure of the steam depot in 1960 can be seen on the far left. *(Alec Swain)*

Passing under one of Temple Meads' numerous gantries containing an array of colour light signals is Warship No D855 *Triumph* with a service for Paddington on 22 May 1963. Originally finished in green livery, this changed to maroon in May 1966. *(Alec Swain)*

Left Top: Another of the hydraulic classes to be found in the west country was the Beyer Peacock Hymek, this example D7045 splitting its short working life of 10 years between Old Oak Common and Bristol Bath Road. It is seen here running light engine along the southern side of Temple Meads returning to depot on 26 June 1963. *(Alec Swain)*

Left Bottom: Perhaps the most unloved of the WR's hydraulic classes, and certainly one of the most unreliable and in many eyes one of the ugliest, were the North British Type 2s. D6350 of this class is seen also on 26 June 1963 near Bath Road Diesel depot where a Warship can be seen on the depot. This locomotive had an even shorter lifespan than the other hydraulic classes putting in just over 6 years before withdrawal in September 1968. This premature exit was not helped by the fact that North British had gone bankrupt in 1962 necessitating utilising withdrawn locomotives for spares in order to keep the rest of the fleet running. The tall depot floodlights are very evident in this view.
(Alec Swain)

Right: We could not leave consideration of the diesel classes without including one of the ubiquitous DMUs that were prevalent on suburban and local stopping services in the Bristol area. Phase One of the Bristol Area diesel scheme commenced in October 1958 when some of the Severn Beach and Henbury line services were turned over to this form of traction. DMUs in the area were a mixture of Derby suburban sets, later Class 116, and both types of Cross Country set, later classes 119 and 120. The example seen here on 26 June 1963 passing the diesel depot is one of the suburban sets sporting its "speed whiskers" as they would become known. *(Alec Swain)*

Turning now to the three motive power depots that served Bristol, we begin at Bath Road which was situated adjacent to Temple Meads, dealing mainly with passenger services. King class No 6015 *King Richard III* is coming off shed on 19 September 1955 ready to work, according to the reporting number 470, the 4:15 (SX) Bristol to Paddington service. Also in view on the left is No 7032 *Denbigh Castle* whose fireman is raking coal forward on the tender. Bath Road was destined to be one of the first WR sheds to be closed to steam locomotives from September 1960. Later rebuilt as a diesel depot it retained one of the turntables formerly used by steam. The diesel depot ceased all operation in September 1995 when Great Western Trains, its last operator, transferred all operations to St. Philips Marsh Traction Depot. *(R C Riley)*

Castle class No 5071 originally named *Clifford Castle* but changed to *Spitfire* from September 1940 (the key month in the Battle of Britain) is seen on Bath Road shed by one of the water columns. The reporting number carried, 825, confirms that the headboard "The Cornishman" is appropriate for this service from Wolverhampton to Penzance. *(Eric Sawford)*

In October 1967 there were six steam locos present at the annual Bath Road Depot open day. These were pannier tank No 1638, No 7029 *Clun Castle*, No 7808 *Cookham Manor*, No 46201 *Princess Elizabeth*, together with a pair of withdrawn Bulleid pacifics Nos. 34013 *Okehampton* and No 34100 *Appledore* which were en route to South Wales for scrapping. There was something of a tradition of withdrawn locos appearing at Bath Road open days for the 1966 event had featured Nos. 48706, 48760 and 80043, all late of Bath (Green Park) shed. There was also a pair of panniers retained for possible preservation on the Clevedon branch. The previous year a motley assortment of steam had included a few refugees from Barrow Road shed the month before it closed whilst in 1970 ex S&D 2-8-0 No 53808 was exhibited en route to Radstock having been rescued from Barry. *(Author)*

The magnificent curved overall roof of Temple Meads looked even more dramatic at night as this atmospheric shot taken one evening in the late 1960s illustrates. A Warship diesel on a passenger working has just arrived at Platform 9 whilst opposite a DMU is in evidence. *(Author)*

Standard Class 5 No 73068 is going well having negotiated the curve leading from Stoke Gifford to Bristol as it passes through Filton Junction with a freight service on an unrecorded date. It was withdrawn from Bath Green Park shed in April 1965 and the station here became unstaffed in 1968 but soldiered on in an ever increasing state of dilapidation until 1996 when it closed. It was replaced by Filton Abbey Wood station serving a nearby large MOD facility. *(Terry Nicholls)*

Hall Class No 4920 *Dumbleton Hall* heads west with a freight service through Parson Street station whose main building was at street level with access to the platforms being by means of the stairs. The main building containing the ticket office, illustrated later on in this volume and dating from the rebuilding of the station in the 1930s, continued to advertise its GW origins until demolished in 1971. The 4-6-0 was latterly allocated to St. Philips Marsh shed from where it was withdrawn at the end of 1965, along with the majority of the WR's remaining stock of steam locomotives. *(Terry Nicholls)*

Left: The sulphurous atmosphere of a steam depot, in this case Barrow Road, is well captured in this view. Amongst the locomotives on view are Black 5 No 44825 and 72XX Class 2-8-0 No 7221. An impressive line of no less than 10 pannier tanks is also evident and in the distance a couple of Barton Hill's tower blocks are under construction. *(Terry Nicholls)*

Right Top: Fishponds on the former Midland exit from the city sees the passage of a Class 45 in 1969 the last year of the route's operation. Following closure, services for the Midlands were routed via Filton, Stoke Gifford and Westerleigh Junction. Fishponds, like other local stations on this route to Bath via Mangotsfield, lost its passenger service on the same day as the closure of the S&D in March 1966. The trackbed is currently heavily utilised as part of the Bristol & Bath Cyclepath. *(Author)*

Right Bottom: With Brunel's Clifton Suspension Bridge in the background, this view of Clifton Bridge station dates from 1967 when the Portishead branchline was still open to coal traffic. The station was subsequently demolished and much of the site is now occupied by the Dog and Mounted section of the local police constabulary, although the single line to Portishead is still in situ and may witness a re-opening to passenger services as part of the MetroWest project. *(Author)*

A scene of yesteryear, when wagonload freight traffic was still important to the railways, is revealed in this panorama of Bristol East Depot as a Class 50 powers past with a parcels train. Coal wagons are evident on the left whilst numerous wagons and vans adorn the many sidings of the up yard. Today the scene is utterly transformed, with the majority of the sidings in the up yard having been removed and much of the site becoming an industrial estate. Bristol East Depot was opened in 1890 to cater for the extra traffic brought about by the opening of the Severn Tunnel. The down yard remained in use as storage for engineering wagons before being lifted at the end of 2004. The following year new sidings were laid on the site of the down yard and the Bristol Steel Terminal opened at a cost of £400,000. This viewpoint also affords a magnificent panorama of the city. *(Author)*

Right Top: On Bristol's sole remaining inner branchline, this is Clifton Down with a single car en route to Avonmouth and Severn Beach. The buildings were, in their day, described as 'commodious and handsome', being built in a modified Gothic style with long curving platforms covered for much of their length with a glass ridge and furrow roof, the remnants of which are still evident in this late 1960s view. Total removal of these canopies came in May 1971. This was the nearest station for the former Bristol Zoo site and many excursions in connection with this attraction were run in past years. The station became unstaffed in July 1967 with Bristol Zoo's Clifton site closing after 186 years in September 2022. This view also shows the site of the once extensive goods yard at Clifton Down to the right with coal being the principal commodity, as witnessed by the remaining sleeper-built coal pens. The signal box seen at the western end of the platform had a twenty-eight-lever frame and continued in operation until November 1970 when the route was largely singled, although a double track loop was retained through the station here and so it remains today. A shopping centre and car park now cover the former goods yard. *(Author)*

Right Bottom: A scene of Bath Road diesel depot full of interest in that it contains a glimpse of one of the Blue Pullman sets which operated between London and Bristol from September 1960 until May 1973, together with a couple of Westerns, a Warship and two Peaks; the one parked next to the snowplough being D62 named *5th Royal Inniskilling Dragoon Guards*, a name not carried until November 1964. There is even a DMU parked at one of the platforms for good measure. *(Terry Nicholls)*

The carriage washer is obviously intriguing the two small boys and their dog looking through the railings in Bristol's Victoria Park which was always a good place to watch the trains go by. Western class hydraulic D1057 *Western Chieftain* is also benefitting from a "wash and brush up" as it heads the stock at the regulation speed, which varied from depot to depot but was generally 3-5mph. *(Terry Nicholls)*

The previously featured Brush Type 4 D1662 *Isambard Kingdom Brunel*, here seen in its attractive two tone green livery, heads away from Temple Meads passing North Somerset Junction with a train of mixed stock that includes one coach still in chocolate and cream livery. An impressive array of semaphores guard the junction which not only gives access to the North Somerset line to Radstock but also allows Temple Meads to be bypassed and serves St. Philips Marsh shed. *(Terry Nicholls)*

Left Top: Class 37 No 37223 waits underneath the trainshed with a Bristol to Cardiff service on 12 June 1983. At this date this English Electric Co-Co locomotive was based at Cardiff's Canton depot and it managed a very creditable 37 years' service before withdrawal came in January 2003 being later scrapped at Sims Metals (now part of Unimetals) in Beeston, Nottinghamshire. *(Arthur Turner)*

Left Bottom: Prototype HST No 252 001 consisting of eight coaches plus two power cars rounds the curve away from Temple Meads on an unrecorded date. This set was constructed at Crewe in 1972 and operated between Paddington and Weston-super-Mare from 1975 until the production units came on stream in quantity. Of interest is the fact that the prototype had buffers whereas the production sets did not. The carriage washer seen in a previous image is evident in the distance. *(Terry Nicholls)*

Right: Snow blankets this scene of Temple Meads taken on 10 February 1985 and captures a cross platform chat between the crews of HST power car No 43141 and Class 31 No 31460. No 43141 was one of 27 sets subsequently moved from Great Western to Abellio Scotrail for refurbishment and can these days be found working between Glasgow/Edinburgh and Aberdeen/Inverness where HSTs are still going strong. *(Arthur Turner)*

Bristol's modern image is represented in this view of Temple Meads taken on 1 June 2024 showing Hitachi Class 800 and Cross Country Voyager units, now the staple motive power for services to London and from the south west to the Midlands and the north respectively. As can be seen from the extensive scaffolding the Grade 1 listed roof is undergoing a major renovation involving grit-blasting in order to remove old paint, dirt and surface contaminants from the metal parts of the structure before being repaired, repainted and reglazed. This multi million pound scheme is part of Network Rail's Bristol Rail Regeneration Programme which aims to transform Bristol into a "World class transport hub" – hopefully fit for the foreseeable future if not for the next 200 years. *(Andrew Royle)*

A closer view of 82A Bath Road steam shed reveals King class No 6004 *King George III* and Standard Class 4 No 75001 in residence on 5 July 1959 - the year before the shed closed to steam. Also in view is part of Grange class No 6863 *Dolhywel Grange*. The brick built depot had ten dead end roads and at this date had an allocation of 82 locomotives including amongst its stud 24 Castles, 13 Halls, 5 Modified Halls and 7 Counties. The large depot clock, assuming of course that it was working, shows this image to have been taken at 2:45pm. *(R C Riley)*

Another King, this time No 6009 *King Charles II*, is also seen on the depot parked next to the coaling stage ramp upon which a couple of coal trucks are positioned and one would hope they were suitably braked as it was quite an incline. This twin-ramp coal stage was to a standard GWR pattern but utilised concrete beams and brick piers to restrict its width. The two railwaymen seated on the far right by the steps to the ramp seem to be having a chuckle about something. *(Eric Sawford)*

The far side of the coaling ramp and tower is seen in this view of a sparkling Castle No 5034 *Corfe Castle* about to receive its next tender full of coal on 5 July 1959. No 5034 was based at Old Oak Common at this time from where it would be withdrawn in September 1962, a victim of the increasing pace of dieselisation on the Western Region. *(R C Riley)*

Taunton based Hall 4-6-0 No 5992 *Horton Hall* rests on Bath Road shed on 31 August 1955 making its own contribution to global warming years before this became such a issue. Like its namesake locomotive Horton Hall, situated on the Northamptonshire/Buckinghamshire border, is now but a memory, the stately late 18th century mansion being sold for development and ultimately demolition in 1936. No 5992 met its fate at Bird's scrapyard at Morriston in September 1965. Also in view is No 7020 *Gloucester Castle* which would precede the Hall to the breakers by a year, being withdrawn from Southall shed in September 1964. *(Eric Sawford)*

Producing nearly as many fumes as did the erstwhile steam depot, no doubt the effect being enhanced by the frosty conditions, this scene at Bath Road Diesel depot was recorded on 10 February 1985. Amongst the locomotives on view are a number of type 33s, of which Eastleigh based No 33004 can be identified, these 'Cromptons' working the through trains from Portsmouth to Cardiff at the time. Also adding to the pollution are examples of classes 47 and 50 together with several 'Peaks'. *(Arthur Turner)*

We now move to the other ex GWR shed at Bristol coded 82B based at St. Philips Marsh which tended to concentrate on freight locomotives. The depot is seen on the right in this view of No 7927 *Willington Hall* passing Marsh Junction on 9 July 1963 with a goods train using this route to avoid travelling via Temple Meads station. This diversion was particularly useful during the summer months when holiday trains from the North and the Midlands to the South West could avoid the congestion at Temple Meads. A new shed was opened here in 1910 as the facilities at Bath Road could not be expanded due to the cramped nature of that site. St. Philips Marsh shed closed to steam in June 1964. *(R C Riley)*

Some of the occupants of the depot on 26 August 1962 include No 6973 *Bricklehampton Hall*, No 7027 *Thornbury Castle*, and 94XX class pannier tank No 8479. Although 7027 escaped the scrapman it subsequently went through many owners but has ultimately, very controversially, been earmarked to donate its boiler for new build 2-8-0 No 4709. *(Henry Priestley)*

28XX class No 3841 is seen at the rear of one of the two roundhouses provided at St. Philips Marsh on an unrecorded date. Carrying an 83A shedplate, this Newton Abbot based 2-8-0 would transfer from there to Severn Tunnel Junction shed in May 1960 and subsequently to Pontypool Road in September 1962 from where it would be withdrawn in March 1964. *(Eric Sawford)*

This image taken within one of the two roundhouses at St. Philips Marsh shed on 9 October 1963 captures the smoky atmosphere of such places. The 65 foot turntable was one of two provided each with 28 roads radiating from it, with each road having its own inspection pit varying in length from 40 – 100 feet. On the left 82B's own No 4093 *Dunster Castle* can be made out in the stygian gloom. A fascinating set of 75 original drawings of the depot are held at 'Steam', the museum of the GWR in Swindon. *(R C Riley)*

The sight of those massive gasholders could only mean that we are now at the third of Bristol's sheds located at Barrow Road. This former LMS facility was photographed by the well known railway photographer George Heiron whose work now forms part of the Transport Treasury archive. Jinty No 47552 is busy shunting the extensive carriage sidings with one rake bearing roofboards "Newcastle – Bristol" whilst steam is rising from several Midland 0-6-0s in the shed yard, just one of which can be positively identified as 4F No 44484. *(George Heiron)*

Royal Scot class No 46117 *Welsh Guardsman* runs light engine tender first from Temple Meads to Barrow Road shed on 5 October 1962 having just worked the down "Devonian", a restaurant car service which ran from Bradford, Leeds, Sheffield and Derby to Exeter, Torquay and Paignton. Engine Shed Sidings signalbox seen on the right was in operation from 1895 until closure came in May 1967. Also on the right are the buildings of the Wagon shops. After 1974 the former running lines here became sidings which were progressively shortened over the years. *(R C Riley)*

This striking portrait of Patriot No 45506 *The Royal Pioneer Corps* was taken at Barrow Road on 5 July 1959. It carries an 82E shedplate reflecting the fact that Barrow Road's Midland shed number of 22A changed to a number in the WR sequence upon transfer of the depot to WR control in 1958. This Fowler 4-6-0 would remain at Barrow Road until withdrawal came in March 1962. Sharing the limelight in this view is 43XX class 2-6-0 No 6376. *(R C Riley)*

A couple of railwaymen observe 4F No 44092 propelling "dead" ex L&YR Aspinall "Pug" No 51218 at Barrow Road shed on 5 October 1962. The 0-4-0ST would go on to a life in preservation on the Keighley & Worth Valley Railway whilst the Fowler 0-6-0 would be withdrawn in September 1964. The depot's coaling tower can be glimpsed behind the arches of the now demolished Barrow Road bridge. *(R C Riley)*

Standard 9F 2-10-0 No 92243 is seen at Barrow Road in this view which although undated by the photographer was probably taken in the last few days of the shed's existence, as No 92243 was allocated here from 7 November 1965 with the shed closing on 30 November that year and being demolished in 1967. The access steps from Barrow Road are visible on the left leading down from the top of the 13 arch viaduct that was also later demolished and that bisected the railway site. *(Milepost)*

Castle class No 7024 *Powis Castle* is in a very woebegone condition at Barrow Road, its glory days well and truly behind it. Devoid of front numberplate with a chalked substitute remarkably it still wears its cabside number and even more amazingly its nameplate. Rust patches and limescale add to its general scruffy appearance but it would be put out of its misery upon withdrawal from Oxley shed in February 1965. *(Terry Nicholls)*

Two views of the large ferro concrete coaling tower provided at Barrow Road along with a water softener and mechanical ash equipment as part of the improvements of the late 1930s. In this first image Ivatt tank No 41240 and 4F No 44272 are in evidence with a coal truck in the process of being hoisted to the top of the tower for emptying. Both locomotives carry a 22A shedplate indicating that this is their home depot. The main line can be seen curving away to the right past the signalbox controlling entry to the shed. *(Eric Sawford)*

Left: This second view shows 28XX class 2-8-0 No 2895 from 88A, which from September 1963 indicated Cardiff East Dock shed, awaiting its next turn of duty. Apparently following closure and demolition of the depot, local Bristolians were rudely awoken one Sunday morning at 6:45am by the demolition of the coaling tower having received little or no advance publicity to warn them. *(Terry Nicholls)*

Right: The graceful lines of Johnson 4-4-0 2P No 40463 dating from 1895 and rebuilt by Fowler with superheater and piston valves is seen in light steam outside the shed shortly before withdrawal in July 1956 after 60 years service. The last example of this elegant class remained in service until 1962. *(Eric Sawford)*

Another Johnson product although rebuilt by Fowler with a Belpaire firebox is this 3F 0-6-0 No 43712 which had worked on the S&D based at Templecombe until transfer to Barrow Road in July 1950. Put into store in December 1959 it was withdrawn the following month. *(Eric Sawford)*

First of its class and built at Derby in 1951 No 73000 graces Barrow Road shed on 5 July 1959 whilst allocated to 41B Sheffield's Grimesthorpe depot. It remained in BR service until March 1968 when it was withdrawn from Patricroft shed. *(R C Riley)*

In this view, No 51212 one of the celebrated Aspinall 0-4-0STs constructed at Horwich Works in 1894 takes centre stage. Its short wheelbase and large dumb buffers were ideal for working Bristol's industrial lines such as that to Avonside Wharf which had curves of very small radius. Firms served by sidings near the wharf included those involved in engineering, paper, lead, distilling and cement. Rail traffic ceased here in 1990. *(Neville Stead Collection)*

Steam and diesel were always uneasy bedfellows as the dirt and grime of a steam shed was not conducive to the requirements of diesel engines. Nonetheless in this view Hymek D7008 from Bath Road depot is on Barrow Road together with 43XX class 2-6-0 No 6327 and a long time 22A allocation since 1947 in the shape of Jubilee No 45690 *Leander*; the latter locomotive of course going on to an illustrious career in preservation following a spell in Woodham's yard at Barry. It is currently out of service at Carnforth awaiting overhaul. *(Peter Pescod)*

Left: The deep rock cutting beyond Bristol's East depot yard sees the passage of 51XX class prairie tank No 4159 with a Keynsham workmens' service on 19 September 1955 passing the East Depot mainline signalbox seen perched high up on the left. *(R C Riley)*

Right: This closer view of the signalbox shows its unusual configuration brought about by the limited space available at ground level allied to the steep sided rock cutting. This elevated box closed in January 1960, replaced by a simpler structure at ground level on the down side of the main line containing a frame with 69 levers which lasted in operation until 1970. *(R C Riley)*

Near North Somerset Junction a rather scruffy Grange No 6831 *Bearley Grange* is in charge of a ballast train, as the position of the headlamps indicates, and has halted under the watchful eyes of several lineside workmen on the girder bridge which crosses the Feeder canal seen underneath at this point. Although this image is undated this 4-6-0 was allocated to St. Philips Marsh shed between June 1958 and January 1962. It is already missing its nameplate and has its cabside number chalked on although its smokebox number remains intact. Like many other examples of the class it ended up at Oxley shed from where it was withdrawn in October 1965, being scrapped along with eight other Granges by Cohen's at their Kettering site. Today Network Rail's Kingsland Road Depot occupies the site of the sidings seen on the left. *(Terry Nicholls)*

Looking eastwards towards London and seen by the substantial girder bridge which crossed the River Avon near Bristol Relief Line Junction this view shows 9F No 92003, reversing light engine towards the signalbox seen in the distance. In the down East Depot yard on the far right the "wasp stripes" of a diesel shunter can just be made out. *(Terry Nicholls)*

4F No 44424 is seen crossing the ex GWR mainline near Narroways Hill Junction with a freight on the former LMS route from Fishponds to Avonmouth in the 1950s prior to its move from Barrow Road shed to Stoke in October 1959. The footbridge seen in the distance is still in situ providing a useful footpath across the valley utilised by the railway here. *(George Heiron)*

This image taken near Narroways Hill Junction on 2 July 1983 shows Class 33 No 33002 hauling a 6 coach Cardiff to Portsmouth service en route to Temple Meads. Also in view is a DMU on one of the four tracks that existed here until two were taken out of use the following year. This short sighted reduction has been made good again by the reinstatement of four tracks between Temple Meads and Filton Abbey Wood at a cost of £33m in 2018. The brick bridge abutments shorn of their girder bridge are all that are left to remind one of the former LMS line from Kingswood Junction near Fishponds to Ashley Hill Junction where it met the Clifton Extension route to Avonmouth, a line that was jointly operated by the GWR and LMS. The connection to the Midland line closed in 1965. *(Arthur Turner)*

A busy scene at Narroways Hill Junction where the, then still double track, Avonmouth line parted company with the mainline. A passenger service is heading down the grade into Bristol hauled by an unidentified Hall class whilst a freight is receiving banking assistance up the incline from 2-6-0 No 6363 under the brooding presence of the gasholders in Glenfrome Road, Eastville. To the right of these metal monsters can just be made out one of the locomotives used to shunt the internal system. *(George Heiron)*

Pictured on 25 August 1969, a few months before closure, St. Anne's Park has already taken on a rather forlorn air following the withdrawal of staff some years before. D7041 heads a Weymouth bound train as a Cross-Country DMU calls at the station. Remaining open longer than many Bristol suburban stations it fell victim to the rationalisation of local services between Bristol and Bath, which also saw the closure of Saltford and the threatened closure of Keynsham. The latter was reprieved and remains open today. *(Bernard Mills)*

Parson Street retains its GW frontage, dating from rebuilding of the station when quadrupling of the tracks took place in 1933, in this 1960s view. All station buildings were demolished in 1971 and metal and glass shelters erected on the platforms and upon the bridge. *(Author)*

Although reduced to single track, Redland station, opened in 1897 on the Clifton Extension Railway, still has much to admire in this early 1970s image. The platform buildings, booking office seen top right, footbridge and vintage signage all add to the charm of this view. The down track was lifted in October 1970 followed by demolition of the down side shelter, footbridge and ticket office in 1973. The main building on the up platform remains today, but no longer in railway use. *(Author)*

The original Brunel station with its impressive hammer beam roof can be seen, beyond the Digby Wyatt 1870s extension, during its period of relegation to a car park. Note the wall mounted signal cabin on the right of this view which remained in use until September 1965. *(Author)*

Left Top: On 18 March 1966 the station pilot at Temple Meads was D4020, later to be renumbered 08 852 under the TOPS scheme. This 0-6-0 shunter was one of almost 1000 examples built over the years by a variety of BR workshops, this one being constructed at Horwich in 1961 and allocated to Bristol from new. It remained here until 1974 when it was transferred to Eastfield depot in Glasgow, from where it was withdrawn in August 1988. Part of Bath Road Loco Yard signalbox is just visible to the left of the Hymek seen in the background. *(Alec Swain)*

Left Bottom: Continuing the shunter theme at Lawrence Hill goods yard, an unidentified Class 03 0-6-0 DM is making up a freight service containing several new commercial vans which the photographer recorded as being destined for Severn Tunnel Junction. This view was taken from the bridge of the former Midland line which crossed GWR tracks here. Also in the yard is an unidentified Class 31 whilst the omnipresent gasholders near Barrow Road can be seen in the background. *(Terry Nicholls)*

Right: We could not leave our look at Bristol's railways without mention of the vast industrial estate of Severnside which over the years has provided much freight traffic for the railways. Its smoking chimneys can be seen in the background to this view of Class 47 No 47108 heading past Hallen Marsh Junction with a train of chemical tankers on 7 June 1979. The line diverging off to the right in the foreground led to Henbury and Filton, whilst that diverging to the right behind the last tanker wagon led to Severn Beach. In the distance are the looming towers of Oldbury Magnox nuclear power station, the last of whose reactors was decommissioned in 2012. Demolition of the reactor buildings and final site clearance is planned for 2096 to 2101 – I wonder what the railway scene will look like then! *(Arthur Turner)*

Until the advent of Parkway station Stapleton Road could be considered to have been Bristol's second station acting as a stopping point for services which did not call at Temple Meads, having used the avoiding line via Dr. Days Bridge Junction. In steam days the station still sported buildings, a water tower, signalbox and an informative noticeboard which proclaimed "Junction for South Wales, Clifton and Avonmouth". Heading towards Temple Meads in this view from the late 1950s is 51XX class 2-6-2T No 5188. Following the decision to reverse Portsmouth to Cardiff services in Temple Meads, service levels at Stapleton Road reduced significantly in the 1960s with goods facilities finishing in 1965 followed by the withdrawal of staff in 1967. *(George Heiron)*

In snowy conditions an HST set powered by Nos 43165 and 43164 pass Stapleton Road with a Plymouth to Leeds service on 12 February 1981. The condition of the girder bridge that the units are seen crossing in this view was advanced as one of the reasons for the reduction from four to two tracks on this section of route in 1984. The viaduct was removed in 2017 as part of the subsequent requadrupling project and replaced with a new structure the following year. A fascinating time lapse video of this replacement is available on Youtube. *(Arthur Turner)*

The open air car park at Bristol's Parkway station has expanded considerably since it opened as its popularity grew and today a multi storey car park is provided. In this view taken some 40 years ago on 14 March 1984 a class 56 No 56033 has charge of a train of stone hoppers whilst an HST set with Nos 43129 and 43130 operates a Swansea to Paddington service. An additional third platform was added to the station in 2007 with a fourth opening in 2018. *(Arthur Turner)*

Coming right up to date we have this view of one of the Hitachi Intercity Express class 802/1 25kV ac overhead/diesel electric 9-car multiple units No 802 102 of the GWR in Brunswick Green livery here operating the 13.29 Swansea - Paddington service approaching Bristol Parkway on 4 May 2019. These new units can operate using bi-mode technology, allowing trains to use both diesel or electric power; electrification from London to South Wales via Bristol Parkway was completed in 2020. However, the current state of electrification on the line from Royal Wootton Bassett to Bristol sees the wires stop short one mile east of Chippenham. The section onwards to Temple Meads has been postponed indefinitely by the Department of Transport due to budgetary constraints. Whilst electrification no doubt pleases the environmental lobby the preponderance of OHL equipment visible in this image does nothing for the aesthetics of the railway scene. *(Hugh Llewlyn, Creative Commons Attribution Share-alike License 2.0)*

This futuristic design of Bristol's Parkway station was completed in 2001 (replacing the original basic building) but will itself be replaced if plans go ahead for a transformation making Parkway into what is euphemistically called a 'living station'. This is a concept that re-thinks what a railway station is – making it "a place that serves local people, communities, businesses and a place where people want to spend time" (Other than waiting for a delayed train of course!). This involves the provision of a community park, a new station building and public space, a nature reserve and green space and a new employment area called 'The Brickworks'. *(Author)*